GOD'S WIGGLY WATER!

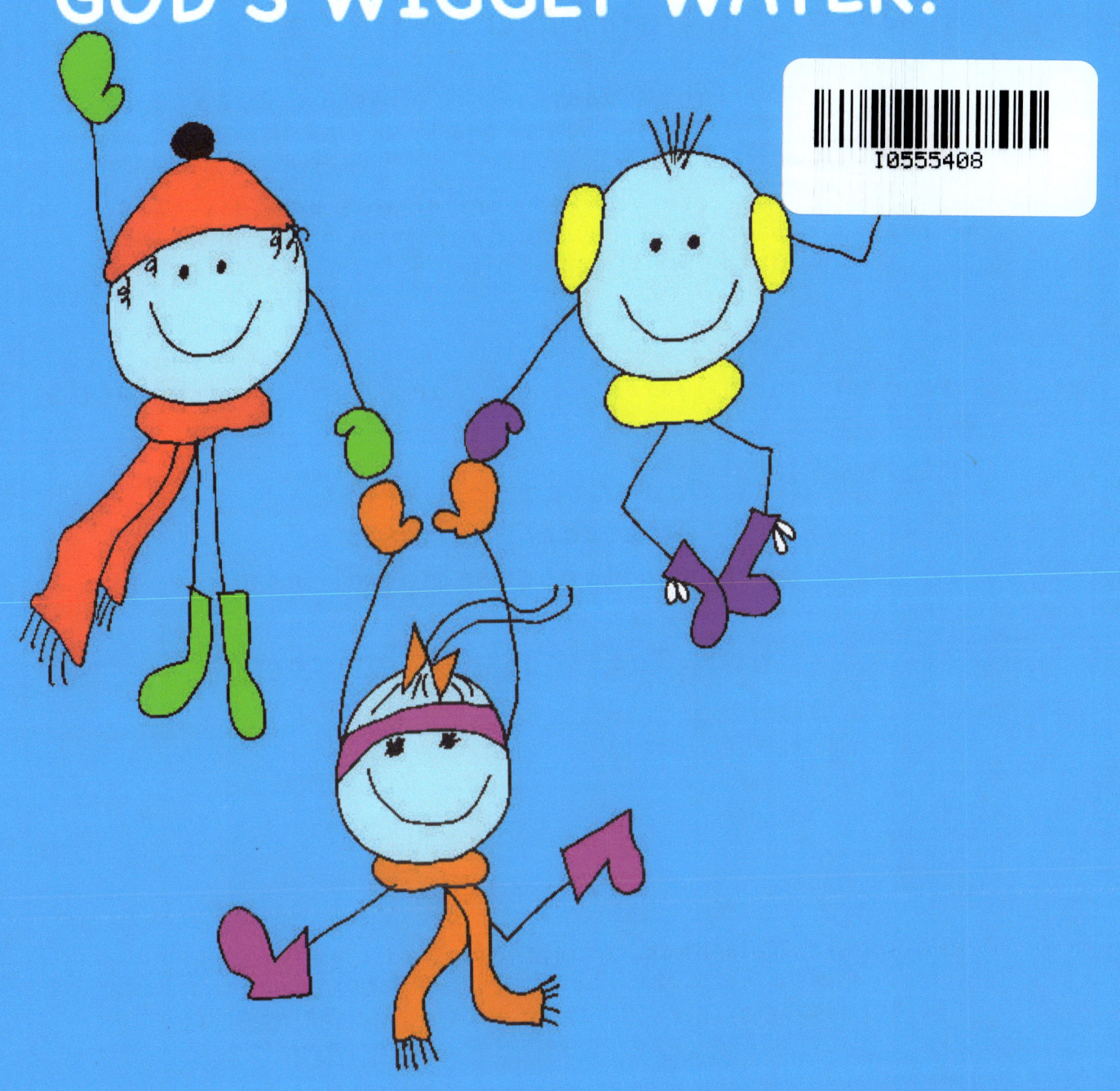

God's Wiggly Water

Book 7 of the God's Cool Creation Book Series

Copyright© Mary Ann Winslow 2023
ISBN: 9798218175177

Written and illustrated by Mary Ann Winslow, PhD

Cool Creation Press
Prescott, Arizona
coolcreationpress@gmail.com

www.Godscoolcreation.com

Disclaimer: This book contains general science information intended to help the reader to better understand basic scientific principles. It is understood that some specific concepts and details have been omitted.

To all those future
scientists out there!
Remember, Jesus loves you!

God's water is amazing! It's everywhere - oceans, rivers, lakes, underground, in the air - and in YOU! In fact, you are over half water! Why is it so special?

For one thing, water is polar – one side of a teeny, tiny molecule (MAHL uh cyool) is positive and one side is negative. So they stick together like tiny best water buddies! This sticking is called cohesion (koh HEE zhun).

Have you ever filled a glass of water until it actually formed a bubble over the top? The water buddies don't want to separate when they're in the air, so they really stick together! That's surface tension (TEN shun), one of God's many designs!

Because water buddies stick together so tightly, there are some insects and lizards that can actually run across the top of water! More surface tension!

More water abilities? They freeze differently than all other liquids in the whole world! When other liquids freeze, they get smaller and heavier, or condense. But the water buddies spread out, get bigger, and get lighter! Who cares?

Fish and marine life do! They would all die if water froze like other liquids! Other liquids freeze from the bottom up, but water freezes from the top down. Ice on top, water below.

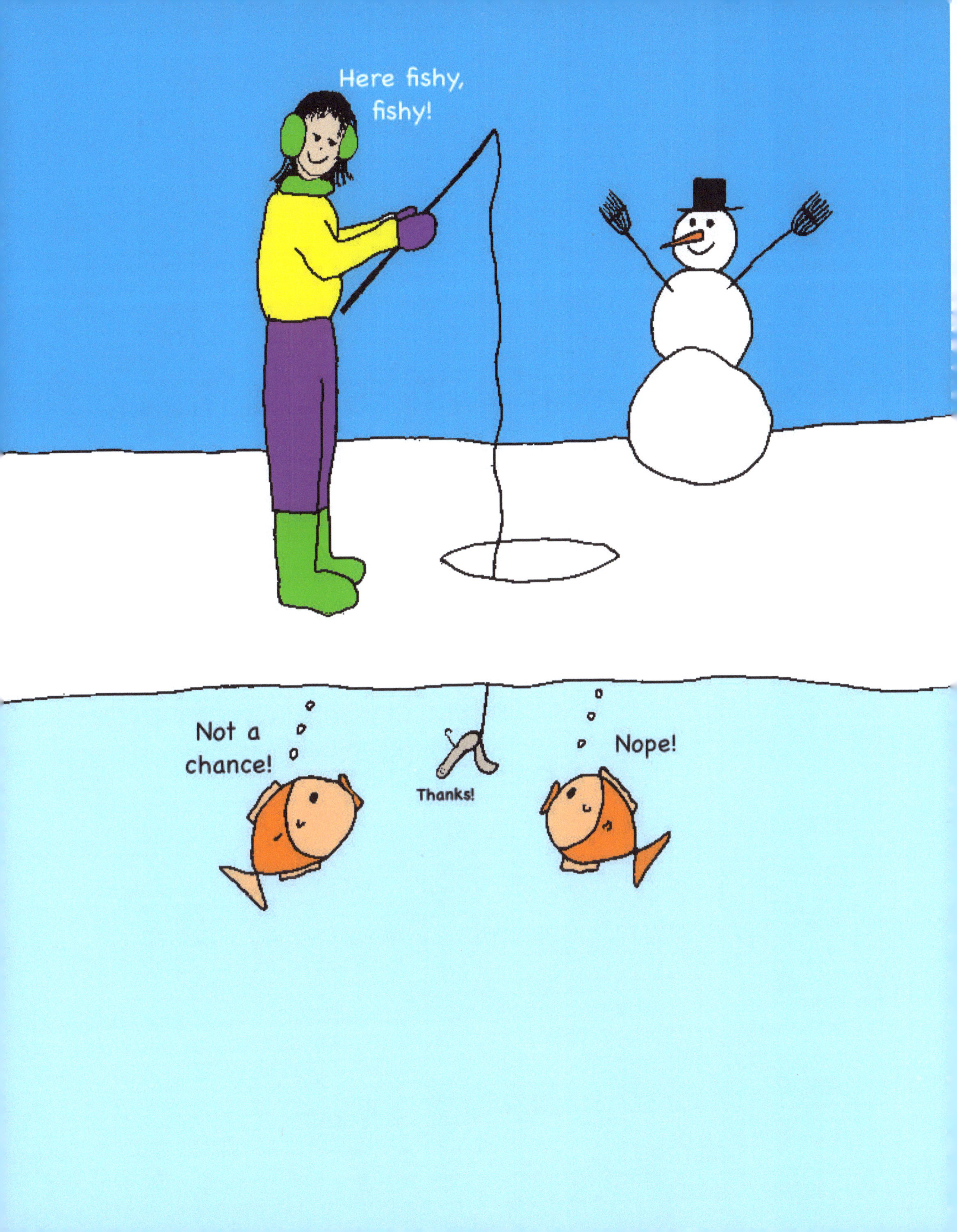

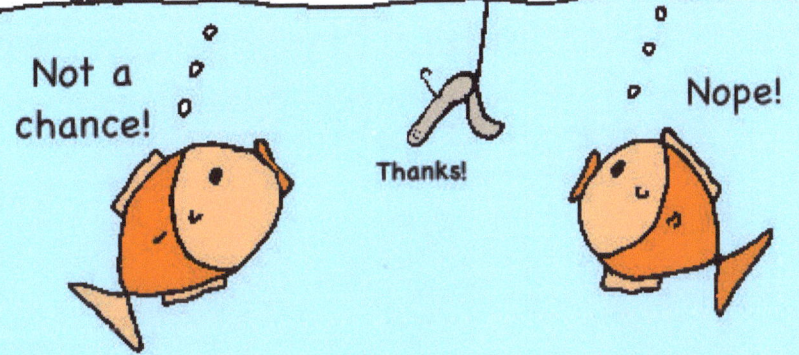

Have you ever walked in a forest or park, and come across a huge rock split down the middle? Our water buddies did that! They seep into tiny cracks in the rock and can actually split the rock from the inside!

Another amazing talent of our water buddies is their traveling abilities! They can travel from tree roots in the ground to the tops of the highest trees! Read God's Cool Creation: Tricky Trees to find out how!

How do the water buddies form clouds? The Sun heats up the water in the ocean and turns the water buddies into vapor (or gas). They're then pulled in the air (evaporation - eee VAP or AY shun).

As they get higher and higher, and colder and colder, water droplets form again on tiny dust floating in the air (condensation - Kahn den SAY shun).

The water droplets come together to form clouds. If they get too crowded and heavy, water buddies fall back to Earth as rain, snow, sleet, or hail (precipitation - pree sip uh TAY shun).

Water buddies sink into lakes, rivers, oceans, and the ground for the plants and trees (infiltration - inn fill TRAY shun), and some sink in aquifers (AKK wuh ferz) - the water buddies gathered below the ground.

The largest aquifer in the United States is the Ogallala (oh guh LAH luh) Aquifer. It's underneath eight states and 175,000 miles of the Great Plains!

Water is more to you than just a drink on a hot day. Sweating keeps you cool when you play hard. And our water buddies are needed in every cell of your body – 30 trillion of them! That's 30,000,000,000,000!

Thank you, Lord!